Solar Photovoltaics for Consumers, Utilities and Investors

Copyright: Steven Magee 2010

Edition 1

Cover Picture: The sun in full view on a clear day in Tucson, AZ, USA

Contents

1. Introduction

Solar photovoltaic technology has now become mainstream and is widely used around the world. All of the items needed to construct a reliable system can now be purchased from many different vendors and standards now exist in the industry to ensure that these different products can be seamlessly integrated together. The future of solar photovoltaics is bright and rapid adoption of the technology is underway.

This book will demonstrate the different choices and components of grid interconnected solar power systems across the different scales of the applications. Solar photovoltaics has longevity with many companies offering twenty five year warranties on their solar modules. When the correct choices are made, excellent systems can be constructed that need minimal maintenance. The book will explore the various factors that affect solar photovoltaics and how to make the correct choices in a relatively young field.

The book covers an overview of the technologies involved for system selection for consumers. For the utilities we will look into the dynamics of site selection and solar module selection for large systems. For investors in solar power there are discussions on what to look for when investing in solar photovoltaic companies. Not all products from companies are created equal and it is important that when purchasing products from or investing in solar photovoltaic companies that you are aware of what to look for.

2. The Basics of Solar Photovoltaics

Solar photovoltaics comes in many different types:

- Monocrystalline
- Polycrystalline
- Thin film
- Other technologies

Although the technology is different between each type, they all do the same thing. All direct current (DC) solar modules generate DC electricity when exposed to sunlight. The listing above is in order of how efficiently each type will convert sunlight into electricity.

A typical direct current (DC) module will have the following electrical ratings on its label:

- Temperature adjustments
- DC Open circuit voltage (Voc)
- DC Maximum power point voltage (Vmpp)
- DC Short circuit current (Isc)
- DC Maximum power point current (Impp)
- DC Rated system voltage

All of these values are given for Standard Test Conditions (STC). Lets look at what each one of these mean:

Standard Test Conditions (STC)

Standard Test Conditions (STC) is how the solar module performs at a temperature of 25 °C, an irradiance of 1000 W/m² with an air mass 1.5 (AM1.5) spectrum. This is a standard test for all solar modules that are manufactured for the USA market that was developed by the photovoltaic industry and the government. It represents an average set of conditions that can be expected at the mid point between North and South of the contiguous forty-eight states during spring and fall with the sun perpendicular to the solar module. San Francisco, California and Wichita, Kansas are near this midpoint of 37 degrees latitude. In Asia Seoul, Korea, is near and in Europe both Sevilla, Spain and Cantania, Italy are near to 37 degrees latitude.

It is important to note that a solar module output will be continuously variable during the year and even during the day. In wintertime it will output less power than its rating and in summertime it will frequently output more power. These electrical ratings are for guidance only and it is where many new photovoltaic designers make mistakes in thinking that the module will never output more power than its rating. It is important that you understand that these solar photovoltaic modules can output far more power than their label states. It can be over fifty percent more and this will need to factored into the system design.

Note for investors: A small number of companies may have built solar power systems that do not fully meet the requirements for the area where they are installed or do not completely adhere to the relevant electrical and building codes. If so, these companies could face a higher level of warranty claims due to an increase in faults on their installations. It is important when investing large sums of money into a solar photovoltaic company that you have confidence in their systems build quality.

Temperature Adjustments

Solar modules are affected by temperature, both hot and cold, and adjustments to the module ratings needs to be made for the operating temperature outside of 25 degrees Celsius. It is important when designing a system that the historical temperature minimum and maximum values are known for the area where the system is to be installed and these adjustments are factored into the design.

Open Circuit Voltage (Voc)

The open circuit voltage rating is how much voltage the module will put out with no load attached. This is an important value in order to design a system. This is the voltage to use when selecting your components and it must be adjusted for the historical minimum and maximum temperatures for the area. If more than one module is connected in series than multiply this temperature adjusted voltage by the number of modules in series to get the total maximum DC voltage of the system.

DC Maximum Power Point

The DC maximum power point is a simple concept. Power is a function of both voltage and current. The maximum power point is obtained when the current and voltage from the module when multiplied together give the maximum power figure. These values will change constantly during the day with the weather conditions. Voltage will remain relatively constant, but current will vary a lot with irradiance. The DC to AC inverter system constantly monitors the power from the solar photovoltaic DC system and automatically keeps the inverter system working at the maximum power value for the given conditions.

DC Maximum Power Point Voltage (Vmpp)

The DC maximum power point voltage (Vmpp) is the operating voltage of the solar module under load. Again this value will change with temperature and irradiance, but should only vary by about twenty percent of the STC rating during the day time.

DC Short Circuit Current (Isc)

The DC short circuit current value is the maximum current that the module will output at Standard Test Conditions if the positive and negative terminals were connected (shorted) together. It is important to note that this value will vary a lot dependent on weather conditions and can be fifty percent larger during summertime.

DC Maximum Power Point Current (Impp)

The DC maximum power point current is the amperage that the solar module will output at standard test conditions in normal operation.   It is important to note that this value will vary a lot dependent on weather conditions and can be over fifty percent larger during summertime.

DC Rated System Voltage

This is very important design value. It is a rating of how many modules can be safely connected together in series. This should never be exceeded when adjusting for the minimum and maximum temperatures of the area that the system is being installed. This value basically limits the number of modules that can be connected in series in the system.

3. Photovoltaics and Weather

The performance of any solar photovoltaic system is dependent on the weather. The main factors that affect the system performance are irradiance, temperature, shade, latitude and how dirty the solar modules are. Let's now explore the effects of the weather in more detail:

Note for investors: Solar photovoltaic companies should avoid locations for large scale projects that may not be well suited for solar photovoltaics. Such areas may have the following environmental attributes:

- Large hail
- High winds (airborne debris)
- Significantly cloudy weather (poor energy yield)
- High level of dirt build up on the solar modules in a short space of time
- High levels of ocean salt spray
- Unstable ground
- Unsuitable area wildlife:-
    - Defecating on the solar modules
    - Gnawing the wiring insulation and associated equipment
    - Using the installed equipment as perches and highways

The result of these conditions is that their systems may not perform well. If they have entered into power purchase agreements (PPA) with energy delivery guarantees with the consumer, then these systems may lose money.

Irradiance

Irradiance is a measure of how much sunlight the solar module is receiving. It is given in watts per meter squared or W/m². Standard Test Conditions (STC) uses a value of 1,000W/m². This value can range from 0W/m² at night through to over 1,500W/m² during a day interspersed with large fluffy clouds. This value of 1,500W/m² is larger than what you would receive in space. The reason why we can get greater values at ground level is due to what is known as the "cloud effect". Normally the sunlight is traveling in a straight line from the sun to our solar module with some atmospheric scattering. However, when clouds are present they can also reflect and can act like lenses to send some extra sunlight onto the solar modules. This extra light is converted into extra energy and this is seen largely as an increase in power from the system. This effect can be a few minutes long in duration when it occurs.

The diagram on the next page demonstrates the "cloud effect".

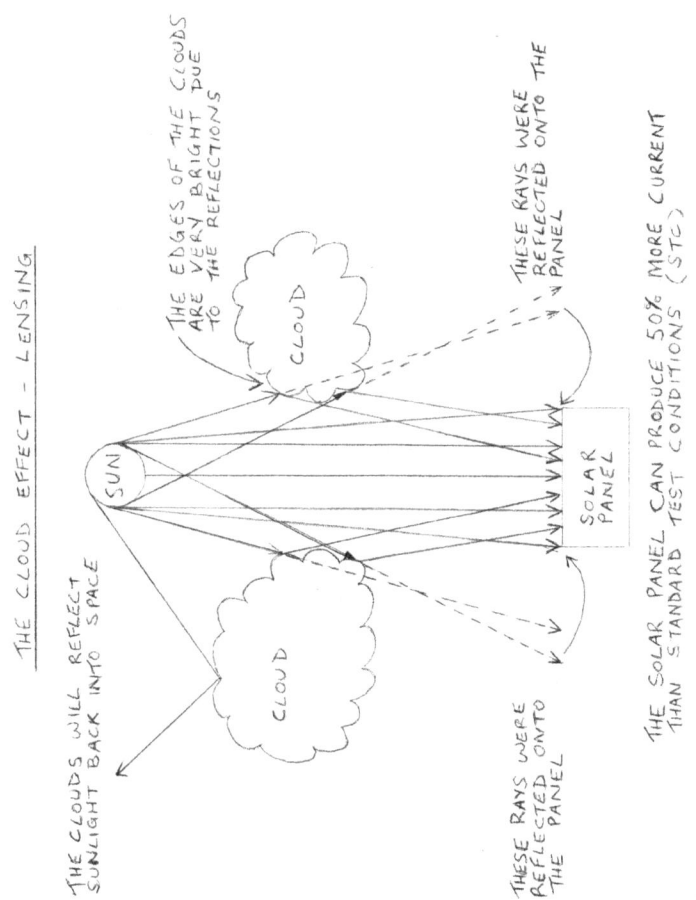

Other effects on irradiance are the snow effect, the lake/ocean effect and the building effect. Snow cover, water, glass covered buildings, reflective painted buildings and roofs can reflect extra sunlight onto the solar power system. If you are installing a system in an area that has any of these, it is important to account for it. Each effect can produce an increase in power output.

In wintertime generally the system will operate at below the standard test conditions values and in summertime it will generally exceed these values. During the design phase of the system you will need to assess where the greatest power need is and perhaps increase the size of the system accordingly if it is in wintertime.

Air Mass

Air mass is a measurement of the the amount of atmosphere that the sunlight has to pass through to get to the ground. It varies with the seasons and also the location on the earth. Within the tropics, air mass will reach its maximum power value of 1 during summertime. Air mass 1 corresponds to the sun being directly overhead, air mass increases as the sun moves from directly overhead down to the horizon.

All USA solar modules are rated for air mass 1.5 which corresponds to a central USA location in Spring and Fall. When in a southerly location you will approach air mass 1 which will increase power output by about 13% from STC in the USA.

Locations that are at or near air mass 1 in Summer time in the USA are all Hawaiian islands, Florida and Texas. Approximately half of the continental USA is located between air mass 1 and air mass 1.5 in Spring and Fall. If you are working on systems that are located in these Southern USA states, you will get more power out of these systems due to a decreased air mass. In summertime the air mass will move closer to 1 in the continental USA.

Note to investors: The systems that yield the most annual energy are located in the desert South West USA and the dry leeward sides of the Hawaiian islands. Hawaii has one of the most expensive rates in the USA for utility electricity and the return on investment (ROI) through energy production is more favorable as such.

Clouds

Clouds come in many forms. An important question is how do clouds affect irradiance on solar power photovoltaic power systems? The list below will help with understanding the effects of clouds on irradiance at air mass 1 (within the tropics in summer time):

– Clear, sunny skies will give approximately 1,130W/m². The transmission characteristics of the atmosphere will vary in clear skies, sometimes being relatively transparent and other times being more opaque and this affects irradiance values. Air quality is a major factor for the transmission of sunlight through the atmosphere. Particulate matter in the atmosphere will reduce the transmission level.

- Thin cirrus will give approximately $1,000W/m^2$. Thin cirrus will give even and relatively stable irradiance levels due to scattering of the light.

- Thick cirrus will give approximately $750W/m^2$.

- Thin clouds will give about $500W/m^2$.

- Thick clouds will give about $250W/m^2$. No shadows on the ground will be present

- Thick clouds with a visibly dark sky will give about $100W/m^2$. No shadows on the ground will be present. You will not be able to see the location of the sun in the sky.

- Tall broken clouds will give surges of about $1,500W/m^2$ and reductions to about $100W/m^2$ of irradiance due to the cloud effect. The rate and length of time for these surges and reductions is dependent on the speed of the clouds passing in front of the sun.

Temperature

Temperature will affect the system to a much lesser extent than irradiance. The cooler the system is below 25 degrees Celsius, the more power it will produce. Correspondingly, the hotter the system is above 25 degrees Celsius, the less power it will produce. Temperature can affect solar photovoltaic systems power output by about twenty percent.

Shade

It is undesirable to shade solar modules as it can significantly affect the performance of the system. When studying the location of where to install a system, always factor in the surroundings for shading effects. Avoid shading with solar photovoltaic power systems.

Wind

Wind will provide cooling to the photovoltaic modules and it is an aid to power production. A breezy location will provide improved performance from the system. When mounting solar modules onto racking, it is good to allow spaces between the solar modules in order to aid with cooling airflow around the modules and also reduce wind resistance. When choosing solar modules and mounting systems, it is important to ensure that they are rated for the wind speed of the area that you are installing them into.

Altitude

A higher altitude location will improve the amount of irradiance that the system will receive, due to less scattering and absorption of the sunlight by the atmosphere. It also acts as a natural cooler of the system which further improves system performance. Generally a high altitude location will have a higher percentage of clearer skies during a year which will give a higher energy yield from the system.

Snow and Ice

Snow and ice should not affect a solar module, other than obscuring its view of the sun. Tracking systems can be affected by this and in some snowy locations it is advisable to park the solar system facing South during these periods. The reflection from the snow will increase the power from the system in Winter time.

Hail

Hail can break solar modules, so it is important to know type of hail that your area can receive. If you get large golf ball size hail, you may not want to install glass solar modules. Solar modules are tested for hail and pass the tests even if the glass module breaks. The test just ensures that the modules remains intact when broken. Glass solar modules are hard to break and normal sized hail should have no effect.

Dirt

Clean solar modules are the desirable configuration for a system. However, dust and dirt will get onto the surface of the modules and will degrade performance by up to 10% on average. Cleaning the modules is very much a function of the location where they are installed and also how dirty they are. Most people will clean on an as needed basis, generally when they are visually very dirty. Always follow the manufacturers instructions for cleaning your particular modules and remember that solar modules are operating

with electricity flowing in them when exposed to light.
Night time cleaning is recommended for safety.

Lightning

Lightning can affect solar modules, especially on large
systems that cover fields. Good equipment grounding is the
way to deal with this threat. A low resistance ground will
generally dissipate lightning away from a solar module that
is struck by lightning. Generally, the damage should be
limited to only the solar module that was struck. If a cable
is struck, then lightning surge arrestors can limit the damage
in the system. These are generally installed in the inverter
and on larger systems, in combiner and re-combiner boxes.
Install lightning protection as recommended by the
manufacturers of the products used in your installation.

Seasons

We have four distinct seasons of Winter, Spring, Summer
and Fall. We can word this another way as Winter Solstice
(December 21), Spring Equinox (March 20), Summer
Solstice (June 21) and Autumn Equinox (September 22).
What does this mean to a solar power system?

- The length of the day
- The angle of the sun (air mass)
- Heating and cooling
- Rain

Winter solstice is the shortest day of the year and summer solstice is the longest day of the year. Spring and Autumn equinoxes are when day time is the same length of time as night time.

Regarding the angle of the sun in the USA, Winter Solstice is when the sun is at the lowest in the sky, or 23.5 degrees below the equator and Summer Solstice is when it is 23.5 degrees above the equator. Spring and Fall equinoxes are when the sun is directly overhead at solar noon at the equator.

For our solar power system, this means that we will produce our largest voltage in wintertime when it is the coldest and we will produce our largest current when it is summertime with peak irradiance.

The changing seasons will affect rainfall and in dry seasons you may want to schedule cleaning to keep the modules in good performance. Rain generally helps to keep the module clean naturally.

There are a number of things to consider with the seasons:

- Spring & Autumn
    - The system will be operating close to standard test conditions (STC) and measured values should be close to that on the solar module label.
    - This is the most favorable time for outdoor working.
- Summer time

- The system will be hot and the DC voltage will be lower than normal.

- The system DC electrical current will be at the highest value for the year.

- Ambient temperatures will be high.

- Heat and dehydration may be a problem for working on the system.

- Wintertime

  - The system DC voltage will be at the highest value for the year

  - It may be too cold to work on the system

  - Frost, ice and snow may be an issue for performing maintenance.

  - Ambient temperatures will be low.

Due Diligence

It is important when designing, operating and maintaining a solar power generation system that you are aware of the annual climatic conditions to expect. Amongst the data that you should have is:

- Historic annual minimum temperature

- Historic annual maximum temperature

- Historic annual maximum wind speed

- Historic annual snow fall depth

- Historic annual hail size

- Historic annual peak irradiance

- Historic monthly irradiance

With these values you will be able to make educated engineering decisions regarding the selection of your system.

4. System Sizing

Sizing your system is a complicated equation of your energy demands and the area that you install your system.

"Net zero" is a common term in the solar industry. This means that your solar photovoltaic system is sized to match your annual consumption of electricity from the grid. In other words you use no electricity from the grid during the year when averaged out over the year. If you are looking for a net zero system, then the first step is to find out how much electricity you use during the year from your electricity bills.

Divide the annual energy consumption by 365 days to get the daily figure.

The next step is to look at the annual solar radiation charts for your area. These can be obtained at the National Renewable Energy Laboratories website at www.nrel.gov. The NREL USA chart for annual average solar data for photovoltaic systems is shown on the next page and was obtained from http://www.nrel.gov/gis/solar.html. I would recommend that you take a look at it in color at the website and download it. You can find out an estimate of how many hours of average daily solar radiation that your area will get from these charts.

Solar photovoltaic system size will be the daily energy consumption divided by the daily solar radiation.

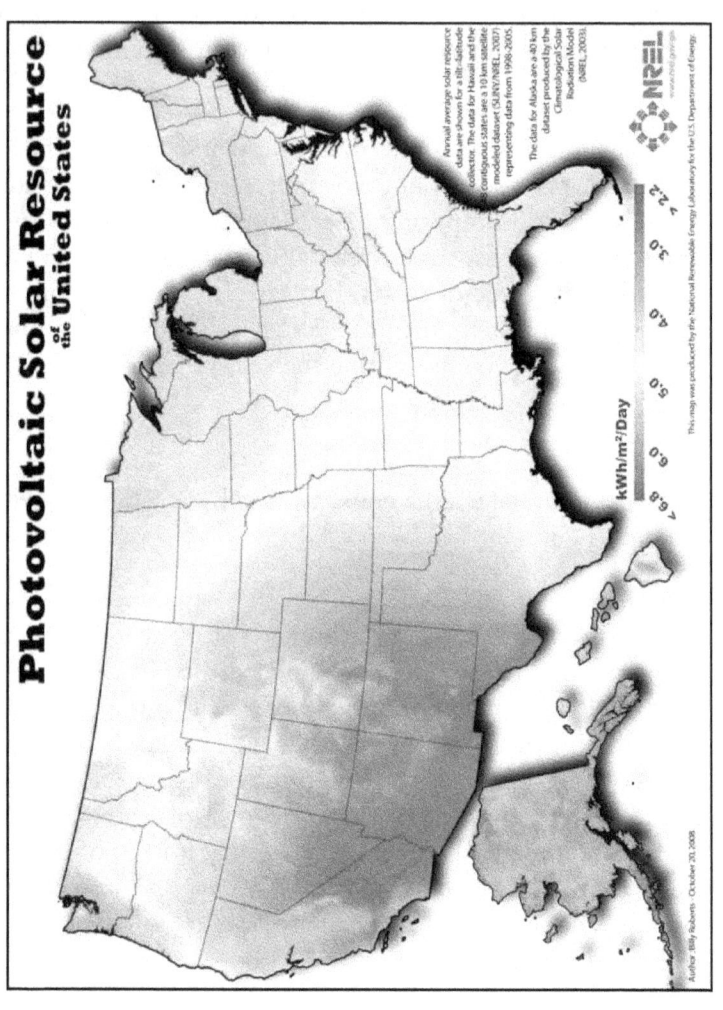

Using the commonly used figure of 0.77 efficiency for the solar power system, adjust the system size figure for the photovoltaic system efficiency losses.

You can adjust for aging effects using a 1% per year of loss of efficiency.

For example:

This particular sample household uses 7,300 kilowatt hours (kWh) of electrical energy per year.

The system that will be installed will be a South facing fixed tilt system that is inclined to match the latitude angle. There is no solar photovoltaic module shading during the day.

Divide 7,300 kWh per year by the number of days per year

= 7,300 / 365

= 20 kWh per day

Looking at the solar radiation charts, this area gets an an annual average of 5 kilowatt hours per square meter per day.

Solar radiation = 5 kWh/m²/Day

We now divide the daily household power requirements by the number of hours of solar energy for this area.

20 kWh / 5

= 4 kW AC solar photovoltaic system size

We need to adjust this for the solar system inefficiencies of 0.77 to get the correct DC system size.

= 4 kW / 0.77

= 5.19 kWp DC solar photovoltaic system size

Since solar power systems do suffer from aging effects by about 1% per year we will factor this in at a 10% loss after 10 years of service and increase the system size accordingly to compensate.

= 5.19 kWp / 0.9

= 5.77 kWp DC

So our system should produce more energy when new and after ten years will hit the net zero figure as it ages.

It is always good practice to oversize solar power systems so that the owner is not disappointed by poor performance. There is nothing worse than a system that does not meet its minimum performance expectations, which in this case is a net zero system.

A further adjustment is required for tracking systems:

A single axis tracker that is inclined at the latitude angle can produce an increase of energy production by about a factor of 1.24. We will reduce the DC system in size accordingly which gives:

= 5.77 kWp DC / 1.24

= 4.65 kWp DC single axis tracker system size

A dual axis tracker can increase the power production by about a factor of 1.3. We will reduce the system size accordingly:

= 5.77 kWp DC / 1.3

= 4.44 kWp DC dual axis tracker system size

You must remember that solar photovoltaics is more of an art than a science and these figures are approximations. It is common for some systems to over perform and others to under perform. It really just depends on your local conditions.

The safest way to quote solar photovoltaic power generation systems is installed DC capacity and installed AC capacity after conversion losses to the point of interconnection with the grid at STC. Annual power production predictions can be unreliable at times and should not be relied on.

Most companies when building a large power plant will build in expansion to the design at the start of the project as an insurance policy in case the system under performs in its location. Generally, a 10% expansion buffer is sensible to use for this purpose. It is wise to oversize the system to exceed the desired annual power production figure and to build it bigger if it under performs once in operation.

The above calculations are based on averaged data from NREL. For a large solar site, they should not be a substitute for a on site study of climatic and system annual performance for at least one year in duration prior to design and construction of a project. It is good to obtain several years of data if possible for a solar photovoltaic site prior to development.

The NREL website has a huge amount of historical solar radiation information on it and it is well worth taking the time to check out their latest information. The NREL data should be the starting point for planning a large solar photovoltaic project.

Now that we have a good understanding of how to perform solar photovoltaic system sizing calculations, let me introduce you to the easy way of doing this. Most solar photovoltaic system designers will use the PVWatt calculators at the NREL website http://www.nrel.gov/rredc/pvwatts/ to size their systems.

5. Solar Modules

Solar modules typically represent the largest investment in the system. Traditionally silicon solar modules have been used, but now many new types of technologies are emerging such as thin film and so on. All solar modules should be tested to the Underwriters Laboratory (UL) 1703 standard in the USA.

In the following chapters we will take a look at each technology and see the needs of each technology when building residential, commercial and utility systems with them. The main solar technologies in use currently are:

- Mono-crystalline silicon wafers (also called single crystal)
- Poly-crystalline silicon wafers (also called multi-crystalline)
- Amorphous silicon thin film
- Cadmium telluride thin film
- CIGS thin film

Silicon is the best understood, the most efficient and has been around for many decades. The newer film type solar modules are less efficient and cheaper to purchase. Unfortunately, currently more thin film modules are needed to generate the same power as mono and poly crystalline modules, and this increases the system physical size and associated support systems such as land purchase, cabling, racking, installation costs, operation and maintenance costs, and so on. Thin film uses less materials to produce it and it really is a film, usually about 1/50 the thickness of a silicon wafer solar cell.

Thin film is a relatively new technology in the solar photovoltaic field. Development of thin film systems was in process in the late nineties and became a commercial product in the last decade. Thin film benefited from a surge in silicon wafer costs and a lack of supply of silicon solar modules in the last decade. Thin film was widely installed during this time. Recently, this silicon supply problem has been rectified and silicon prices have fallen sharply and this has been hurting the thin film solar module manufacturers due to the superior efficiency of silicon wafers.

A further problem with thin film is that its longevity is still being assessed. Some thin film manufacturers offer a 25 year warranty but cannot demonstrate commercial product that is of that age. It will come with time, but financing companies are reluctant to finance thin film due to the higher risks that aging poses for these systems. In time, the various thin film solar modules are expected to be proven to have a similar reliability as silicon wafer solar modules.

Thin film systems are generally marketed as the solar panel of choice in cloudy locations due to their improved performance in low light conditions when compared to silicon. Thin film is less efficient than silicon and is currently between 7 and 10 percent in the commercial products. As such more solar panels are needed when compared to silicon systems to generate the same energy at STC.

Generally any decision on which technology to use is driven by market rates for each type of technology, aesthetics and personal preference. Solar modules are a commodity and their prices can fluctuate rapidly. When designing a system, it may be built from any of the above mentioned solar modules.

Roofing materials are now being manufactured with solar photovoltaics integrated into them. This is a great idea and will bring down the installation costs on new construction. The products are relatively new and it will be interesting to watch this area develop. Generally this technology is based on thin film.

Be careful when using manufacturers data sheets for their solar modules. The module power value generally has percentage tolerances for negative and positive adjustments to this figure. This is a reflection of the imprecision of the solar photovoltaic cell manufacturing processes. Due to this it is common for individual strings of solar modules to either over or under perform in the system. For example, the cloud effect combined with a 10% high performing string of solar modules could produce 165% more than the rated STC current from the string. If your system under performs it may be due to having the bulk of the supplied modules performing at lower than normal power ratings. At least one major manufacturer does not list any negative adjustments on its solar photovoltaic module data sheets, only positive adjustments. This greatly helps with meeting the system performance requirements. It would be good to see other manufacturers follow their lead.

There are some manufacturers that offer 25 year warranties but do not have any product that old. You will need to assess if this is an issue in your decision to use their products. A good question would be to ask yourself if you think that the manufacturer will be in existence in 25 years time to honor the warranty.

Note to investors: Solar module manufacturers have emerged and have product in the market place without a long history of a successful and long lasting product. Their solar module longevity is unproven over twenty-five years. The older and well established solar module manufacturers have developed excellent reputations in solar photovoltaics due to the proven longevity of their products. UL 1703 solar photovoltaic module testing helps assure that these younger products will last the full twenty five years of their warranty period.

We will look into the physical sizes of three systems for each solar module type. These will be:

- 5 kWp DC typical residential system
- 250 kWp DC typical commercial system
- 10 MWp DC typical utility system

By comparing the system sizes and components you will be able to see the advantages of each of the different mainstream technologies.

We will not look at the newer solar photovoltaic technologies due to them being in their infancy. Their longevity is currently unknown and most large financing companies will not finance these newer technologies in large installations.

The picture on the following page demonstrates the string concept of solar modules. Solar modules are first wired in series, this is called "stringing", to increase their DC voltage output. Then many strings are connected in parallel to increase the current output of the solar system. Raising the voltage and current is needed for the inverter system to work properly and to increase the efficiency of the system.

Let us now take a look at each of these technologies and see the differences between typical systems of the same size that are built using the different technologies.

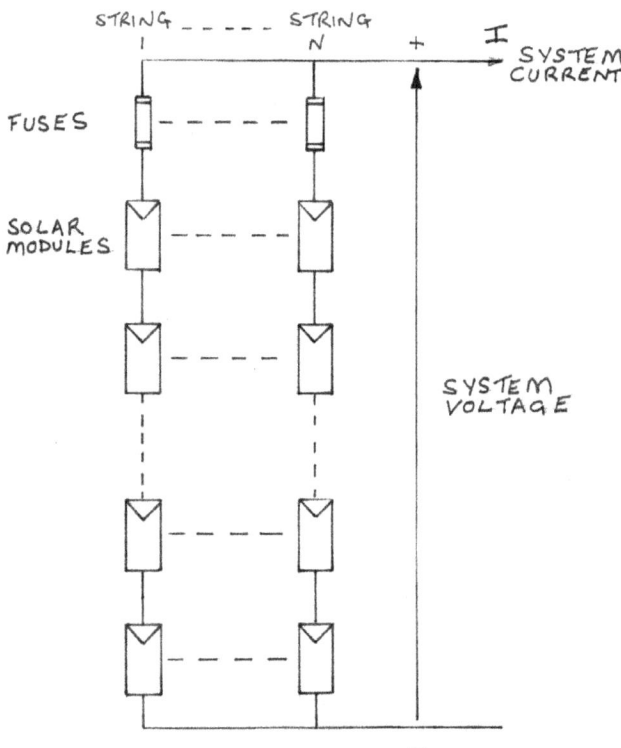

SOLAR MODULE STRINGS

6. Mono-crystalline Modules

Mono-crystalline modules are the highest efficiency solar modules available in the general market. What does this mean? Basically an installation constructed with these modules will use less support equipment as these systems use the least amount of solar modules for a given energy requirement. This advantage comes at a cost, as these modules are also the most expensive.

They are made out of a high grade silicon wafers and are very reliable. The mono-crystalline solar cells look black to the eye. Mono-crystalline solar modules have been around for many decades and are very well understood and commonly come with 25 year warranties. Most major manufacturers can show you modules that exceed this age and that are still functioning well today.

Mono-crystalline from most manufacturers is in the 14 to 18 percent efficiency rating. A few have actually manage to exceed 18 percent with over 20 percent being achieved commercially.

Typical Mono-crystalline Solar Module Specifications

- 600V maximum system voltage
- 235W power rating +10% / -5%
- 37V open circuit (Voc)
- 8.6A short circuit (Isc)
- 30V maximum power point (Vmpp)

- 7.84A maximum power point (Impp)
- -0.351% voltage (Voc) temperature coefficient
- +0.053% Current (Isc) temperature coefficient
- 15A maximum fuse
- Length 1.64m / 64.6"
- Width 0.994m / 39.1"
- 1.63m² / 17.53 sq ft module surface area

System Calculations

Looking at these dimensional values we get a square meter power of:

W/m^2 = rated power / surface area

$= 235W / 1.63m^2$

$= 144.17 \ W/m^2$

We dived this value by ten to get the module efficiency per square meter:

Module Efficiency $= 14.417\%$ per m^2

Assuming a lowest ambient temperature of -10 °C for this area, we need to calculate the maximum voltage for this solar module:

Vmax = (Difference from STC temperature x Voc temperature coefficient x Voc) + Voc

= ((-35) x (-0.351%) x 37V) + 37V

= 41.55V maximum adjusted solar module voltage

Maximum number of solar modules in the strings:

= Rated system voltage / maximum adjusted solar module voltage

= 600V / 41.55

= 14 solar modules maximum in the strings

Wattage of maximum individual string:

= # modules x module wattage

= 14 x 235W

= 3,290W per string

<u>5kWp DC System</u>

Number of modules required in the 5kWp system:

= System wattage / module wattage

= 5,000W / 235W

= 22 Modules

strings = # system modules / # modules in string

= 22 / 14

= 2 strings, each of 11 modules

5kWp system surface area:

= # modules x module surface area

= 22 x 17.53 sq ft

= 385.66 sq ft module surface area

<u>250 kWp DC system</u>

Number of modules required in the 250 kWp system:

= System wattage / module wattage

= 250,000W / 235W

= 1,064 Modules

strings = # system modules / # modules in string

= 1,064 / 14

= 76 strings

Number of 12 input combiner boxes required:

= # strings / # combiner box inputs

= 76 / 12

= 7 combiner boxes required

250kWp system surface area:

= # modules x module surface area

= 1,064 x 17.53 sq ft

= 18,651.92 sq ft module surface area

10 MWp DC system

Number of modules required in the 10MWp system:

= system wattage / module wattage

= 10,000,000W / 235W

= 42,554 rounded up to 42,560 Modules

strings = # system modules / # modules in string

= 42,560 / 14

= 3,040 strings

Number of 12 input combiner boxes required:

= # strings / # combiner box inputs

= 3,040 / 12

= 254 combiner boxes required

10MWp system surface area:

= # modules x module surface area

= 42,560 x 17.53 sq ft

= 746,076.8 sq ft module surface area

Let's make an assumption that when constructed and allowing for spacing between the modules for shading effects that this system will cover 100 acres. We will compare the sizes of our other technologies to this to develop the size of land needed for each technology type.

7. Poly-crystalline Modules

Poly-crystalline solar modules are made from lower grade and therefore cheaper silicon. As such, their efficiency is less than mono-crystalline, but not by that much. A poly-crystalline solar module has a typical conversion efficiency of 13 percent to 16 percent. Their purchase price reflects this lower efficiency and the dollar per watt figure is typically less than mono-crystalline. Poly-crystalline solar cells typically look a metallic blue color to the eye. They are visually a very attractive module to the eye.

Poly-crystalline solar modules have been around for many decades and are a very stable and proven technology.

Typical Poly-crystalline Solar Module Specifications

- 600V maximum system voltage
- 215W power rating +5W / 0W
- 33.2V open circuit (Voc)
- 8.78A short circuit (Isc)
- 26.6V maximum power point (Vmpp)
- 8.09A maximum power point (Impp)
- -0.361% voltage (Voc) temperature coefficient
- +0.06% Current (Isc) temperature coefficient
- 15A maximum fuse
- Length 1.5m / 59.1"
- Width 0.99m / 39"

- 1.485m² / 16 sq ft module surface area

System Calculations

Looking at these dimensional values we get a square meter power of:

W/m^2 = rated power / surface area

= 215W / 1.485m²

= 144.78 W/m^2

We dived this value by ten to get the module efficiency per square meter:

Module Efficiency = 14.478% per m²

Assuming a lowest ambient temperature of -10 °C for this solar site, we need to calculate the maximum voltage for this solar module:

Vmax = (Difference from STC temperature x Voc temperature coefficient x Voc) + Voc

= ((-35) x (-0.361%) x 33.2V) + 33.2V

= 37.395V maximum adjusted solar module voltage

Maximum number of solar modules in the strings:

= Rated system voltage / maximum adjusted solar module voltage

= 600V / 37.395

= 16 solar modules maximum in the strings

Wattage of maximum individual string:

= # modules x module wattage

= 16 x 215W

= 3,440W per string

<u>5kWp DC System</u>

Number of modules required in the 5kWp system:

= System wattage / module wattage

= 5,000W / 215W

= 24 Modules

\# strings = \# system modules / \# modules in string

= 24 / 16

= 2 strings, each of 12 modules

5kWp system surface area:

= \# modules x module surface area

= 24 x 16 sq ft

= 384 sq ft module surface area

<u>250 kWp DC system</u>

Number of modules required in the 250 kWp system:

= System wattage / module wattage

= 250,000W / 215W

= 1,168 Modules

\# strings = \# system modules / \# modules in string

= 1,168 / 16

= 73 strings

Number of 12 input combiner boxes required:

= # strings / # combiner box inputs

= 73 / 12

= 7 combiner boxes required

250kWp system surface area:

= # modules x module surface area

= 1,168 x 16 sq ft

= 18,688 sq ft module surface area

<u>10 MWp DC system</u>

Number of modules required in the 10MWp system:

= system wattage / module wattage

= 10,000,000W / 215W

= 46,512 modules

strings = # system modules / # modules in string

= 46,512 / 16

= 2,907 strings

Number of 12 input combiner boxes required:

= # strings / # combiner box inputs

= 2,907 / 12

= 243 combiner boxes required

10MWp system surface area:

= # modules x module surface area

= 42,560 x 17.53 sq ft

= 744,192 sq ft module surface area

Comparing this to our mono-crystalline system we get:

Land size = 100 acres x poly surface area / mono surface area

= 100 x 744,192 / 746,078.8

= 99.74 acres

As we can see, there is no significant difference between the area for the mono and poly crystalline modules in this application.

8. Amorphous Silicon (a-Si)/ Microcrystalline Silicon Thin Film

Commonly referred to as the calculator solar cell, due to its use in small solar powered consumer devices. This material has been around for many decades and its properties are well known and understood.

<u>Typical Amorphous Silicon / Microcrystalline Silicon Solar Module Specifications</u>

- 600V maximum system voltage

- 135W power rating

- 249V open circuit (Voc)

- 0.87A short circuit (Isc)

- 188V maximum power point (Vmpp)

- 0.72A maximum power point (Impp)

- -0.3% voltage (Voc) temperature coefficient

- +0.07% Current (Isc) temperature coefficient

- 2A Fuse

- Length 1.409m / 55.5"

- Width 1.009m / 39.7"

- 1.422m^2 / 15.29 sq ft module surface area

<u>System Calculations</u>

Looking at these dimensional values we get a square meter power of:

W/m^2 = rated power / surface area

= 135W / 1.422m²

= 94.94 W/m²

We divide this value by ten to get the module efficiency per square meter:

Module efficiency = 9.494% per m²

Assuming a lowest ambient temperature of -10 °C for this solar site, we need to calculate the maximum voltage for this solar module:

Vmax = (Difference from STC temperature x Voc temperature coefficient x Voc) + Voc

= ((-35) x (-0.3%) x 249V) + 249V

= 275.2V maximum adjusted solar module voltage

Maximum number of solar modules in the strings:

= Rated system voltage / maximum adjusted solar module voltage

= 600V / 275.2V

= 2 solar modules maximum in the strings

Wattage of maximum individual string:

= # modules x module wattage

= 2 x 135W

= 270W per string

<u>5kWp DC System</u>

Number of modules required in the 5kWp system:

= System wattage / module wattage

= 5,000W / 135W

= 37 rounded to 38 modules

strings = # system modules / # modules in string

= 38 / 2

= 19 strings

5kWp system surface area:

= # modules x module surface area

= 38 x 15.29 sq ft

= 581.02 sq ft module surface area

250 kWp DC system

Number of modules required in the 250 kWp system:

= System wattage / module wattage

= 250,000W / 135W

= 1,852 Modules

strings = # system modules / # modules in string

= 1,852 / 2

= 926 strings

Number of 12 input combiner boxes required:

= # strings / # combiner box inputs

= 926 / 12

= 78 combiner boxes required

250kWp system surface area:

= # modules x module surface area

= 1852 x 15.29 sq ft

= 28,317.08 sq ft module surface area

<u>10 MWp DC system</u>

Number of modules required in the 10MWp system:

= system wattage / module wattage

= 10,000,000W / 135W

= 74,075 rounded up to 74,076 modules

\# strings = \# system modules / \# modules in string

= 74,076 / 2

= 37,038 strings

Number of 12 input combiner boxes required:

= \# strings / \# combiner box inputs

= 37,038 / 12

= 3,087 combiner boxes required

10MWp system surface area:

= \# modules x module surface area

= 74,075 x 15.29 sq ft

= 1,132,606.75 sq ft module surface area

Comparing this to our mono-crystalline system we get:

Land size = 100 acres x silicon film surface area / mono surface area

= 100 x 1,132,606.75 / 746,078.8

= 151.818 acres

As we can see, there is a significant difference between the land requirements of the silicon crystalline modules and the silicon film in this application. At this point you would be wise to calculate the initial land costs, installation costs, and the expected operation and maintenance costs for the lifetime of the system.

9. Cadmium Telluride (CdTe) Thin Film

This is the market leader in thin film. It is a coating process that is applied directly to the glass. A nice feature of it is that it does not need bypass diodes, so the solar module junction box is much simpler. This is a relatively young technology and is in widespread use and well understood. It has better high temperature performance than the other technologies.

<u>Typical Cadmium Telluride Solar Module Specifications</u>

- 600V maximum system voltage

- 70W power rating

- 88V open circuit (Voc)

- 1.23A short circuit (Isc)

- 65.5V maximum power point (Vmpp)

- 1.07A maximum power point (Impp)

- -0.25% voltage (Voc) high temperature coefficient

- -0.2% voltage (Voc) low temperature coefficient

- +0.04% Current (Isc) temperature coefficient

- 10A Fuse

- Length 1.2m / 47.244"

- Width 0.6m / 23.622"

- 0.72m^2 / 7.745 sq ft module surface area

<u>System Calculations</u>

Looking at these dimensional values we get a square meter power of:

W/m^2 = rated power / surface area

= 70W / 0.72m^2

= 97.22 W/m^2

We divide this value by ten to get the module efficiency per square meter:

Module Efficiency = 9.722% per m^2

Assuming a lowest ambient temperature of -10 °C for this solar site, we need to calculate the maximum voltage for this solar module:

Vmax = (Difference from STC temperature x Voc temperature coefficient x Voc) + Voc

= ((-35) x (-0.2%) x 88V) + 88V

= 94.16V maximum adjusted solar module voltage

Maximum number of solar modules in the strings:

= Rated system voltage / maximum adjusted solar module voltage

= 600V / 94.16V

= 6 solar modules maximum in the strings

Wattage of maximum individual string:

= # modules x module wattage

= 6 x 70W

= 420W per string

<u>5kWp DC System</u>

Number of modules required in the 5kWp system:

= System wattage / module wattage

= 5,000W / 70W

= 72 modules

strings = # system modules / # modules in string

= 72 / 6

= 12 strings

5kWp system surface area:

= # modules x module surface area

= 72 x 7.745 sq ft

= 557.64 sq ft module surface area

<u>250 kWp DC system</u>

Number of modules required in the 250 kWp system:

= System wattage / module wattage

= 250,000W / 70W

= 3,572 Modules

strings = # system modules / # modules in string

= 3,572 / 6

= 596 strings

Number of 12 input combiner boxes required:

= # strings / # combiner box inputs

= 596 / 12

= 50 combiner boxes required

250kWp system surface area:

= # modules x module surface area

= 3572 x 7.745 sq ft

= 27,665.14 sq ft module surface area

<u>10 MWp DC system</u>

Number of modules required in the 10MWp system:

= system wattage / module wattage

= 10,000,000W / 70W

= 142,858 modules

strings = # system modules / # modules in string

= 142,858 / 6

= 23,810 strings

Number of 12 input combiner boxes required:

= # strings / # combiner box inputs

= 23,810 / 12

= 1,985 combiner boxes required

10MWp system surface area:

= # modules x module surface area

= 74,075 x 15.29 sq ft

= 1,106,427.47 sq ft module surface area

Comparing this to our mono-crystalline system we get:

Land size = 100 acres x CdTe film surface area / mono surface area

= 100 x 1,106,427.47 / 746,076.8 acres

= 148.3 acres

As we can see, there is significant difference in land use between the silicon crystalline modules and the CdTe film in this application. At this point you would be wise to calculate the initial land costs, installation costs, and the expected operation and maintenance costs for the lifetime of the system.

10. CIGS Thin Film

CIGS comprises of copper, indium, gallium, and selenium in its manufacturing process, hence its name. CIGS thin film looks a streaky greenish black to the eye due to the bath process that is used during the manufacture of the product. CIGS was developed in the nineties and has been a relatively new entrant to the solar photovoltaic field. It is a vacuum deposition process.

Typical CIGS Thin Film Solar Panel Specifications

- 600V maximum system voltage

- 150W power rating

- 96V open circuit (Voc)

- 2.5A short circuit (Isc)

- 70.5V maximum power point (Vmpp)

- 2.15A maximum power point (Impp)

- -0.38% voltage (Voc) temperature coefficient

- -0.06% Current (Isc) temperature coefficient

- 12A Fuse

- Length 1.82m / 71.65"

- Width 1.08m / 42.52"

- 1.97m^2 / 20.65 sq ft module surface area

System Calculations

Looking at these dimensional values we get a square meter power of:

W/m^2 = rated power / surface area

= 150W / 1.97m²

= 76.14 W/m²

We divide this value by ten to get the module efficiency per square meter:

Module efficiency = 7.61% per m²

Assuming a lowest ambient temperature of -10 °C for this solar site, we need to calculate the maximum voltage for this solar module:

Vmax = (Difference from STC temperature x Voc temperature coefficient x Voc) + Voc

= ((-35) x (-0.38%) x 96V) + 96V

= 108.77V maximum adjusted solar module voltage

Maximum number of solar modules in the strings:

= Rated system voltage / maximum adjusted solar module voltage

= 600V / 108.77V

= 5 solar modules maximum in the strings

Wattage of maximum individual string:

= # modules x module wattage

= 5 x 150W

= 750W per string

5kWp DC System

Number of modules required in the 5kWp system:

= System wattage / module wattage

= 5,000W / 150W

= 34 rounded up to 35 modules

strings = # system modules / # modules in string

= 35 / 5

= 7 strings

5kWp system surface area:

= # modules x module surface area

= 35 x 20.65 sq ft

= 722.75 sq ft module surface area

250 kWp DC system

Number of modules required in the 250 kWp system:

= System wattage / module wattage

= 250,000W / 150W

= 1,667 rounded up to 1,670 Modules

strings = # system modules / # modules in string

= 1,667 / 5

= 334 strings

Number of 12 input combiner boxes required:

= # strings / # combiner box inputs

= 334 / 12

= 28 combiner boxes required

250kWp system surface area:

= # modules x module surface area

= 1,667 x 20.65 sq ft

= 34,423.55 sq ft module surface area

<u>10 MWp DC system</u>

Number of modules required in the 10MWp system:

= system wattage / module wattage

= 10,000,000W / 150W

= 66,667 rounded up to 66,670 modules

\# strings = \# system modules / \# modules in string

= 66,670 / 5

= 13,334 strings

Number of 12 input combiner boxes required:

= \# strings / \# combiner box inputs

= 13,334 / 12

= 1,112 combiner boxes required

10MWp system surface area:

= \# modules x module surface area

= 66,667 x 20.65 sq ft

= 1,376,673.55 sq ft module surface area

Comparing this to our mono-crystalline system we get:

Land size = 100 acres x CIGS film surface area / mono surface area

= 100 x 1,376,673.55 / 746,076.8

= 184.52 acres

As we can see, there is significant difference between the land area for silicon crystalline modules and the CIGS film in this application. At this point you would be wise to calculate the initial land costs, installation costs and the expected operation and maintenance costs for the lifetime of the system.

11. Solar Module Summary

For residential installations, the supporting infrastructure costs vary little between the different technology types for installation, operation and maintenance.

This dynamic starts to change as the system gets larger with commercial and utility installations. As we can see, just an decrease in conversion efficiency of 1% can add acres of land use and supporting infrastructure to a 10 MW project. The industry is constantly trying to find more efficiency out of their products due to this.

If land use is a significant project expense or is limited, then you probably should be using the more efficient technologies.

Some areas have very cheap land and if this is the case, the more important figures to look at are the supporting infrastructure costs for a larger installation and the ongoing annual operation and maintenance costs for a larger power generation facility. In the dry deserts of the South West USA, these operation and maintenance costs are generally quite low.

Thin film has better performance in low light conditions when compared to the silicon wafer technologies and this may be a factor to consider if the installation area is frequently cloudy.

If the failure rate is the same across all technologies, then the you will be replacing more equipment during the year on a less efficient installation during routine maintenance of the system. Manufacturers appear to be reluctant to release solar module failure rates for their solar module products and on any project, I have become accustomed to assuming a 1% annual failure rate in absence of this information. Unusual and unexpected severe weather conditions may increase this failure rate. Severe weather insurance is recommended to cover these events.

Operation and maintenance costs will be costly for some projects due to:

- Poor site selection
- Poor equipment selection
- Poor wildlife assessment
- Poor build quality

Performing due diligence at the start of any project will keep these problems under control. Ensuring that a highly skilled solar photovoltaic consultant is overseeing large projects from conception through to design and onto construction and commissioning will help to prevent these problems from occurring.

Site selection is probably the most important part of any large solar photovoltaic system project and can be the determining factor in a projects overall success. On any large project it is recommended to install a small version of the proposed system for the first year prior to constructing the main project so that an assessment can be made of how the system will perform in that location and design changes can be incorporated into the main project once all the risks are known. In the large scale solar photovoltaics field it is the tortoise that is generally more successful, try not to be the hare.

Be wary of locations close to the ocean and take the necessary precautions with equipment selection so that high winds, salt spray and corrosion don't become a problem. Salt will deposit itself onto the surface of the modules and if it is excessive, you will be frequently washing the solar photovoltaic modules. Expect the operation and maintenance costs to be higher when installing next to the ocean.

Frame-less solar photovoltaic modules are becoming available and these have better self cleaning properties during rains when compared to framed modules. Frame-less modules are not as strong as framed modules and you will need to assess if this is a problem for your installation or not. Some frame-less modules do not require a ground connection and this can make a significant saving on a large project.

The table on the next page shows the different technologies.

Module Type	Module Wattage	Efficiency	Modules in Strings	# Modules	10 MW Area	Combiner Boxes
Mono	235W	14.42%	14	42,560	100 acres	254
Poly	215W	14.48%	16	46,512	99.74 acres	243
a-Si	135W	9.49%	2	74,076	151.8 acres	3,087
CdTe	70W	9.722%	6	142,858	148.3 acres	1,985
CIGS	150W	7.61%	5	66,667	184.5 acres	1,112

12. Inverters

Inverters convert the DC solar photovoltaic module power into AC power that can be put into the grid. Grid tied inverters come in many different types and sizes. This book is dedicated to grid tie inverters and this is what we will consider. All residential and commercial solar photovoltaic grid tie inverters should comply with UL1741 in the USA. This ensures reliability throughout the various climate zones that the USA has to offer.

There is a lot of discussion as to where inverter technology is going. There are cost savings with using the largest inverter available for large utility installations, but this comes at the expense of poorer MPPT tracking and long DC cable runs which cause power losses. The DC currents are very high in these systems at several thousand amps. Large solar photovoltaic inverter technology (>250 kW) is relatively new in the industry and may not be as reliable as the established small (<10kW) and medium (10kW to 250kW) sized inverters.

AC solar modules are starting to appear and this is at the smallest end of the inverter spectrum. Each solar module has it's own inverter attached to it, commonly called a "micro" inverter. This gives great MPPT performance. This is the ideal configuration for systems that may suffer from:

- Shading
- Have solar modules facing in different directions from each other

- Heavy dirt accumulation

Long term, as the "micro" inverter technology reduces in price I believe that this may become the standard in the industry, even on utility scale installations. The installation is simplified and removes DC from the circuit which can be problematic due to arcing effects which can cause equipment to burn out when it occurs.

Looking at today's technology I currently recommend conventional inverter systems for all residential, commercial and utility scale installations. I do not recommend inverters over 250 kW as the DC currents and power flowing through these systems is very large and may cause problems in the solar DC circuit.

Micro inverters can be useful where there is nowhere to mount the traditional inverter on a domestic installation, or if the owner considers an inverter box ugly. The micro inverters can be mounted to the solar panels and hidden from view.

The utility scale inverter systems are still in flux in the USA and currently there is no UL standard for these systems. This will come and for now, the USA utility scale inverter industry is to be regarded as experimental.

When connecting strings to the inverter, do not mix and match solar panel types and connect them to a single inverter, this is a recipe for disaster. If you desire to install the different technology types on the same site, then each solar module technology must have its own inverter system dedicated to it.

Module mismatch in strings is also an issue for the inverter MPPT. Do not connect different wattage modules together in the same string otherwise you will promote the hot solar cell effect in the lower wattage modules, as the higher wattage modules will try and push their current through the lower wattage ones. This will waste power. Each individual solar string needs to have identical wattage modules in it.

Foreign manufactured inverter systems may arrive with the inverter set for a different voltage or frequency of AC electricity. Always check that the inverter system matches the grid system that you are connecting into.

Note to investors: Any inverter system that does not have a Underwriters Laboratory (UL) listing or equivalent USA specific testing certification may be a potential risk in the USA market.

13. Switchgear

Switchgear is widely used in solar photovoltaic installations. A complete line of switchgear has been developed for this application. Specifically, combiner and recombiner boxes are now listed amongst most manufacturers product lines.

Older switchgear was not constructed with solar power generation in mind and may only be rated for electricity flow in one direction. If so, they will have markings on the electrical terminals that say either "load" or "supply".

Solar photovoltaic systems feature a AC bi-directional current flow into the grid that reverses twice per day. During the day time the solar power system will generate AC energy into the grid and during the night, the current will reverse and it will consume a small amount of power from the grid. On cloudy days this current reversal will occur frequently during the day time on net zero systems. As such, all switchgear in the AC circuit on a grid connected solar power system will need to be rated for bi-directional current flow. Due to their bi-directional operation, these electrical systems have no "load" or "supply" markings on the electrical terminals.

Investor note: The switchgear manufacturers who have added solar photovoltaic compatible products to their product line will be best placed to to take advantage of the increasing solar photovoltaic demand in the world. There will be a demand for solar photovoltaic compatible fuse boards for residential and commercial applications as consumers change out their old fuse

boards with new ones at the same time as installing solar photovoltaic power systems. It is expected that all fuse boards used in new home construction in the future will be rated for solar photovoltaic use.

14. Distribution

Distribution is a feature of the larger commercial installations and the utility installations. Basically distribution covers the medium voltage application which is normally at 24 kV in the USA.

Solar photovoltaic systems feature a bi-directional current flow that reverses twice per day. During the day time the solar power system will generate energy into the grid and during the night, the current will reverse and it will consume a small amount of power from the grid. As such, all distribution equipment in a solar photovoltaic power system will need to be rated for bi-directional current flow.

Investor note: As the distribution equipment suppliers certify their product lines for use with bi-directional current flow for solar photovoltaics, the ones who will be best placed to take advantage of this will be the suppliers who have working systems installed in the field and can demonstrate safe and reliable systems.

15. Progressing Photovoltaic Adoption

There are some simple steps that can be taken to progress easy adoption of solar photovoltaics for the consumer:

- New buildings could have a solar photovoltaic system installed during the planning and construction phase. The installation costs would be very low during this phase.

- When constructing new subdivisions, install smart grid technologies into them so that the utility can monitor the solar photovoltaic energy feeding into the grid.

- Every new home constructed could have a solar photovoltaic rated (bi-directional) fuse board installed as standard with spare breaker locations labeled for a future solar photovoltaic grid connected system.

- Install the infrastructure for a future solar photovoltaic system during construction, both the electrical wiring and mounting system roof supports. The consumer only needs to buy the solar panels, mounting system and inverter later.

- Build houses with a true south facing roof that is ready to have solar installed onto it.

16. Hybrid Power Plants

Future power generation plants will most likely be built on thousands of acres of land and be hybrid power plants. They may be a mix of fossil fuel/biofuel, solar photovoltaic, solar thermal, geothermal, hydro, wave, wind, nuclear, tidal, or other technologies depending on their location.

A typical conventional power plant has about a gigawatt (1,000,000,000W) of generation capacity. To put that into perspective with solar photovoltaics, the majority of large solar photovoltaic power plants in the world today are just a megawatt (1,000,000W) in size, or one thousandth of the generation capacity of a conventional power plant. The gigawatt solar photovoltaic power plant will arrive in the near future. When it does it will most likely have the following specifications:

- 10,000 acres of land
- Fixed tilt solar panels
- No moving parts
- DC solar panels
- 1,000 to 4,444,445 inverter systems
- 200 to 1,000 dry transformers
- 3,333,333 to 14,285,714 solar panels, depending on the solar photovoltaic technology chosen.
- A full time staff of approximately 25 people to perform the routine operation and maintenance of the system.

As you can see, it will be very large and need a dedicated team of engineers and technicians to keep it in operation. It would cost in the region of 5 to 6 billion dollars at 2010 prices. (And I would like to be the consulting engineer!)

It would be built next to a conventional fueled power plant and used to offset fuel consumption of that power plant.. The large physical size of the solar photovoltaic power plant will average out the broken cloud effect.

17. Potential Industry Problems

There may be problems in an industry that is relatively young. Some of these could be:

- Poor solar module construction
- Poor solar power system construction
- Inverter failures
- Poor warranty coverage
- Business survival

Lets look at each of these and the implications of them

Poor Solar Module Construction

Companies may produce solar module products with various faults that show up many years later. Some of these faults could be:

- Module de-lamination
- Solar cell mismatch
- Poor framing
- Poor mounting system

The results of any of the above conditions in the product may cause premature failure before the 25 year warranty expires. This could prove very costly for companies and their reputations if it occurs in their products.

Poor Solar Power System Construction

Poor construction could involve any of the following problems:

- Incorrect module mounting
- Incorrect system wind speed ratings
- Incorrect grounding system
- Incorrect cable sizing
- Incorrect component selection
- Incorrect foundation systems
- Poor site selection

Any of the above could prove costly for a company during the system warranty period which is typically five years for a complete system or up to 20 years for a power purchase agreement (PPA).

Inverter Failures

The inverter is a consumable item and should be expected to need replacing after about ten years of service. There is a lot of protection built into an inverter and it is difficult to damage an inverter.

There has been a rush to develop inverter systems by the large electrical companies that are moving into the solar field. New to market inverter systems that have no long term history are a higher risk purchase. Just because the company has good heritage, it does not mean that their new inverter product has the same.

The companies with a long solar inverter history are familiar with the requirements of a solar photovoltaic inverter and are well versed with the design requirements of these products. For the newer companies and products, the UL1741 inverter testing standard assures reliability of the products.

Poor Warranty Coverage

Not all warranties are created equal. A company with a good reputation that has been in the solar field for many decades has worked hard to make their products reliable and to honor their warranty. A consumer can have a lot of confidence in these warranties.

Many of the newer companies also have excellent warranty coverage and it will be interesting to see how they develop, particularly with regards to warranty replacement and repair. These companies will be defined in the future by the reliability and service of their products and warranties.

Business Survival

There are many new companies appearing in the solar field with various product offerings. When purchasing a product that comes with a twenty or twenty five year warranty, then it is important to have the confidence that your supplier will still be in business over this time period. The newer companies will gain this consumer confidence over time.

Note to investors: The coming years will most likely see a shake out in the industry. Some companies will disappear, some will be taken over and others will grow into strong and very large organizations. A careful assessment of a company's potential future growth should be made prior to investing into them. This should be done by a solar photovoltaic industry consultant and should include an inspection of sample installations for design and build quality, a review of warranty claims made against the company and references obtained from previous customers.

18. Summary

Solar photovoltaics is very reliable technology. The field is set for growth and systems should become commonplace with the support of the government incentives. When designed and installed correctly on a good site, solar photovoltaic power systems will reliably produce energy for many years with minimal maintenance.

In the future it is quite possible that every new home constructed will come with a solar photovoltaic power system as standard. Installing solar photovoltaics during the construction phase of a home is the most cost effective way to adopt the technology.

Many companies have solar photovoltaic roofing products on the market and in development to meet the future needs of modern, self contained buildings. Net zero is probably the future of green construction, where a building will produce its own energy needs with an annual energy use of zero from the utility grid.

With the development and adoption of AC solar module technologies, solar photovoltaics will not need any specialized knowledge to install it and any electrician will be able to work with it. AC solar photovoltaic modules will be just like an appliance that will plug into the house electrical system. Indeed, in the future you may well purchase your solar photovoltaic modules at the appliance shop.

The smart grid is in development to allow these distributed generation systems to be effectively managed by the utilities. The electrical grid will be quite different in the future with power generation taking place on the majority of consumer premises with the utility grid supporting it when sufficient power cannot be generated due to climatic conditions.

The utilities will continue to build large solar photovoltaic power plants and they will keep getting larger with each generation of plant. The utilities are currently learning how to construct solar photovoltaic systems that work well with their existing infrastructure. Today's utility systems are paving the way towards the very reliable solar photovoltaic power generation systems of the future.

It is an exciting time in the solar photovoltaic industry and I hope that this book has given you an overview of where the technology is today and where it may be heading in the future.

19. References

- NFPA National Electrical Code 2005
- NFPA National Electrical Code 2008
- IEEE/ANSI-C2-2007 National Electric Safety Code
- Occupational Safety and Health Administration www.osha.gov
- National Renewable Energy Laboratories www.nrel.gov
- United States Department of Energy www.energy.gov
- Solar Photovoltaic Training for Residential, Commercial and Utility Systems by Steven Magee
- Solar Photovoltaic Design for Residential, Commercial and Utility Systems by Steven Magee
- Solar Photovoltaic Operation and Maintenance for Residential, Commercial and Utility Systems by Steven Magee

20. Author Contact

Steven Magee,

3618 S. Desert Lantern Road,

Tucson,

AZ 85735

USA

I hope that you found the book informative and please let me know about any questions or comments about the book.

I am a consultant on new solar photovoltaic projects, solar photovoltaic troubleshooting, solar photovoltaic training, and solar photovoltaic investing for financial companies. Please feel free to contact me for any help or assistance in these areas.

You may find my other books useful:

- Solar Photovoltaic Training for Residential, Commercial and Utility Systems

- Solar Photovoltaic Design for Residential, Commercial and Utility Systems

- Solar Photovoltaic Operation and Maintenance for Residential, Commercial and Utility Systems